Digestive Health

Barbara Wexler, MPH

Copyright © 2009 by Barbara Wexler, MPH

All rights reserved. No part of this publication may be reproduced, stored in a retrieval system, or transmitted in any form without the prior written permission of the copyright owner.

For permissions, ordering information, or bulk quantity discounts, contact:
Woodland Publishing, Salt Lake City, Utah
Visit our Web site: www.woodlandpublishing.com
Toll-free number: (800) 777-BOOK

The information in this book is for educational purposes only and is not recommended as a means of diagnosing or treating an illness. All matters concerning physical and mental health should be supervised by a health practitioner knowledgeable in treating that particular condition. Neither the publisher nor the author directly or indirectly dispenses medical advice, nor do they prescribe any remedies or assume any responsibility for those who choose to treat themselves.

Cataloging-in-Publication data is available from the Library of Congress.

ISBN-13: 978-1-58054-116-9

Printed in the United States of America

Contents

What Is the Digestive Tract?	5
A New View of the Immune System	6
Sidebar: When Good Germs Go Bad	6
The Invisible Web of Life	7
A World Within a World	9
When Inside Is Really Outside	10
What Is GALT?	11
Sidebar: The Ecology of Immunity	12
Who's Minding the House?	12
Favoring Our Friends	13
Intestinal Conditions and Diseases	16
When Is an Allergen Not an Allergen?	18
Acidity and Digestive Health	19
Supporting Digestive Health	20
Probiotics	20
Prebiotics	23
Digestive Enzymes	24
Some Food for Thought	25
Natural Support for Digestion	26
Test Your GI IQ	26
References	28

What Is the Digestive Tract?

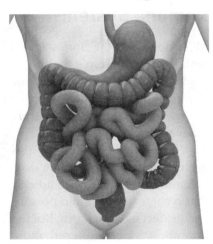

The digestive tract includes multiple organs, including the esophagus, the stomach, the small intestine, the colon and the rectum.

The digestive tract, also known as the alimentary canal, the gastrointestinal tract (GI tract) or the gut (an old Anglo-Saxon term), is a long (nearly 30 feet), winding organ in the body. The digestive tract includes multiple organs involved in the process of digestion: the esophagus, the stomach, the small intestine, the colon and the rectum.

Three quarters of the immune system is found in the digestive tract. When the intestinal frontier of the immune system is healthy and working well, it plays an amazing role in maintaining and restoring health. It distinguishes nutrients from pathogens (disease-causing agents) and helps to diminish the intensity of allergies and food sensitivities. The

When Good Germs Go Bad

You may have heard about a "killer germ" called *Escherichia coli*, or E. coli. But E. coli is a usually harmless bacterium that lives by the billions in our intestines. What makes it so deadly?

Bacteria exchange genes—little bits of DNA that hold chemical recipes—as a kind of language. Unlike the DNA in cells, the DNA in bacteria floats around freely in loops called replicons. Most of a bacterium's DNA is contained in a big loop called a large replicon, but bacteria can also have small replicons that hold as little as a single chemical recipe.

One thing bacteria do well is cook up new recipes that help them survive. If we subject bacteria to antibiotics, they quickly respond by finding a chemical solution to protect themselves—a kind of anti-antibiotic. They tuck this recipe into a small replicon called a plasmid and carry it around for protection from antibiotics.

Imagine a really noxious bacterium that can cause serious illness. This bacterium is probably dangerous because it contains genes that manufacture chemicals that poison the human body. If the human population encountered this bacterium before and treated it with antibiotics, it's likely that the bacterium has developed anti-antibiotic genes along with its poison genes. Such bacteria are antibiotic resistant.

These poisonous bacteria have a chance to mix with the friendly bacteria in our bodies, and two of
(continued on page 8)

digestive tract also plays a key role in transporting dietary fat into the blood.

The portion of the immune system in the digestive tract (known as gut-associated lymphoid tissue, or GALT) enables people to resist infection and to be resilient to health challenges and threats. We are completely unaware of GALT when it is performing its functions. Each day, GALT helps us draw nutrients from food, regulates immune responses and protects us from harmful substances.

A New View of the Immune System

Most books on the immune system, from children's books to graduate school texts, have one thing in common—they describe the immune system as an army ready to defend the body.

There is some truth to this notion. The immune system does contain some cellular "soldiers" that recognize germs and poisons and protect us from harm. But the immune system is bigger, more awe-inspiring and, frankly, more miraculous than the biological army analogy implies.

Modern biology is discovering that no organism on earth—from the lowliest single-cell bacteria to

A New View of the Immune System • 7

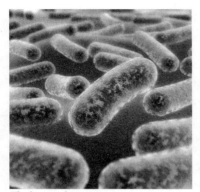

The human digestive system contains beneficial bacteria.

humans, dolphins, rosebushes and everything else—has ever been immune to its neighbors. This lack of immunity turns out to be one of the most powerful driving forces in evolution. Throughout history, cells have merged, mixed and matched their individual capabilities in cellular communities called consortia to become better adapted to their natural environments. Looking closely, we've discovered subtle evolutionary processes that Charles Darwin, the 19th century English naturalist, never dreamed of—processes that are based more on cooperation, communication and community building than on competition, chance and survival of the fittest.

As we've carefully examined living things to find out how they really work, we've discovered that organisms are in a constant state of communication. They send chemical and energetic signals to mark their turf and coordinate their activities. Organisms even learn new capabilities by trading genetic codes, much like kids swapping baseball cards. *(See sidebar: "When Good Germs Go Bad.")*

Living things do not try to achieve an impossible state of *immunity*; rather, they work to maintain their *integrity*—their ability to live and thrive and reproduce in a world full of other things, each trying to do the same.

The lack of immunity at the cellular level is so profound that scientists are now hard-pressed to say exactly where one species leaves off and another begins. The following is an example from nature—specifically, the interaction between trees and mushrooms in a forest.

The Invisible Web of Life

Go to a forest and take a good look. You'll notice that some of the trees live along the edge of the forest while others are deep in the woods. Clearings throughout the forest create extra edges. Trees use sunlight to make cellular energy, and they use that energy to produce nutrients critical for growth and survival. Naturally, the

(continued from page 6)
these bacteria may get close enough to communicate chemically. In effect, the poisonous bacterium says, "Hey! I've got this useful little plasmid here that destroys antibiotics. Do you want it?" And the friendly bacterium says, "Cool! Send it on over."

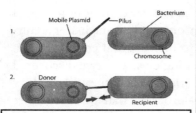

Schematic drawing of bacterial conjunction. 1-Donor cell produces pilus, 2-Pilus attaches to recepient cell, brings the cells together. Adapted from original drawing by Mike Jones for Wikipedia.

At this point, the first bacterium grows a temporary, funnel-like organ called a pilus. It docks the pilus into the other bacterium and makes a copy of its antibiotic resistance plasmid, which it transmits through the pilus. Voila! The second bacterium now has the same antibiotic resistance.

But suppose the first bacterium has another plasmid—one that generates a poisonous chemical. In some cases, this second plasmid can also slip through the funnel and into the other bacterium. Suddenly, a previously friendly bacterium, which our immune system recognizes as an ally, now carries the recipe for a potentially deadly poison. It has not only become a bad germ, it has become a bad germ that our system recognizes as a friend.

(continued on page 10)

trees along the edges of the forest receive extra sunlight, because they're more exposed. And because these trees receive more sunlight than the trees in the shade, they make more nutrients.

However, under the earth, billions of tiny mushroom threads, called mycelia, wrap around the roots of the trees and link them into a giant web. The extra nutrients produced by the trees in the sun are carried along these threads to feed the trees in the shade, much like our circulatory system carries oxygen and nutrients throughout the body. In effect, the mycelia act as a giant circulatory system, uniting many trees into a single living entity.

There's more. These mycelia (the vast underground mushrooms that pop up through the soil) also need certain nutrients. Because they lack the capacity to make some of these nutrients, they rely on enzymes within the roots of the trees to feed them. So while the mycelia are acting as a circulatory system for the trees, the trees are acting as a digestive system for the mycelia.

If you didn't know this, could you fully understand the nature of a tree or a mushroom? Trees and mushrooms may seem like separate entities, but they are connected in a way similar to, for example, an arm and our liver— two completely different body parts working together to make a single person. When we look closely, we start to understand the deep interconnections that define all living things.

The more we understand how living things connect and communicate, the more we understand how ecosystems embody complex webs of life. The British biologist James Lovelock argues that the Earth is a single, conscious living being he calls Gaia (pronounced GUY-uh). Lovelock describes how the earth works to maintain the integrity of its metabolism at a planetary level through the exchange of energy and nutrients between the atmosphere, the oceans and the life on the face of the earth.

A World Within a World

The digestive tract is filled with living organisms that behave much like webs of underground fibers linking trees in a forest. The body contains trillions of friendly, single-cell bacteria that help digest and assimilate nutrients and maintain a healthy relationship with the outside world. The symbiotic relationships we maintain with friendly bacteria have evolved over millions of years. Even though these bacteria are independent living things (if they were taken out of our bodies they could survive on their own) they are an essential part of the body. Friendly bacteria are an important part of the immune system.

More than 100 trillion bacteria from more than 400 different species live in the intestines, primarily in the colon. The combined weight of these bacteria is between three and four pounds. Termed intestinal flora or friendly flora, these beneficial bacteria are essential for healthy digestion and immunity.

Beneficial bacteria synthesize such vitamins as vitamin A, vitamin B_{12}, biotin and vitamin K. The bacteria break down toxins and prevent overgrowth of harmful microorganisms. They increase the bioavailability (the extent to which nutrients are available for the body to use) of minerals such as calcium, copper, iron and magnesium. Friendly flora also stimulate the immune system and produce short chain fatty acids that are essential for healthy colon cells.

When Inside Is Really Outside

It seems surprising, but the digestive tract is technically part of the *surface* of the body. Although the digestive tract is located inside the body, it's actually an outer surface that's folded inward, like the hole of a donut or the inner part of a soda straw. The things we put into our mouths don't go directly to the true inner compartments of the body. Instead, they pass through a series of staging areas on the inwardly folded surface of the digestive system. At different stages along the journey, specialized tissues and organs decide which parts of the food stream can be broken down into useable nutrients and taken through the barriers into the body's *actual* insides. This not only gives the body the chance to separate nutrients from wastes but also the chance to identify and reject some of the toxins and germs that travel with our food.

Though it seems to be on the inside, the digestive tract is like the hole of a donut.

Because of its exposure to the outer world, the digestive tract must not only digest and transport materials taken in, it must analyze the materials, as well. First, the body must determine if the material is food. If the material is not food, the body must determine if it is a threat. The immune system also must recognize the body's own tissues so it does not try to digest them or attack them. Seventy-five percent of the immune system lies along the digestive tract.

When we are healthy, the immune system functions silently in the digestive tract. When something goes wrong, however, we may experience a wide range of symptoms: indigestion, gas, bloating, food allergies and sensitivi-

---

(continued from page 8)

This is what may have happened with Shigella, a pathogenic bacterium that generates a potentially lethal toxin. It is believed that sick cows, treated with antibiotics, accidentally passed the Shiga-toxin plasmid to otherwise harmless E. coli bacteria along with their antibiotic resistance plasmids.

The new "killer" version of E. coli carried the plasmid for the Shiga-toxin to other cattle. As a result, cow meat, if raw or undercooked, could now pass the infection on to humans.

In the U.S. more than 70,000 cases of infection have been reported, and as of 2007, 61 deaths were attributed to STEC—Shiga-Toxin E. Coli bacteria.

ties, increased susceptibility to infections, or emotional and mental disturbances such as attention deficit disorder and depression.

What Is GALT?

GALT (gut-associated lymphoid tissue) is a network of immune system cells (lymphoid tissue) located throughout the digestive tract. The components of GALT include the following:

Tonsils. The tonsils are located on both sides of the back of the throat and are visible when you open your mouth and say, "ahh." Although their precise role in the immune system has not yet been pinpointed (until the mid-twentieth century, doctors routinely removed infected tonsils), the tonsils are thought to protect against infection, particularly respiratory infection.

Tonsils are at the front line of intestinal immunity.

Adenoids. Similar to the tonsils, the adenoids are spongy masses of tissue at the back of the nose on both sides of the upper part of the throat. Together, the tonsils and adenoids form an open ring of tissue in the throat called Waldeyer's ring. The adenoids swell when they become infected, which can cause difficulties in breathing.

Peyer's patches. Along the walls of the lowest portion of the small intestine are clusters of lymphoid tissue known as Peyer's patches. These patches are a few centimeters long and contain white blood cells that perform vital immune functions.

GALT also includes clusters of lymphoid tissue in the esophagus, stomach, appendix and large intestine, as well as lymphoid cells and plasma cells in the connective tissue *(lamina propria)* that lines the digestive tract.

GALT performs two primary immune functions—defense and repair. In its defense role, GALT detects invaders called antigens, which provoke the immune system to produce antibodies, and haptens, which cause an immune response but not one that produces

The Ecology of Immunity

A large amount of communication in the body takes place through chemical messengers. For years, biologists have tried to separate these messengers into different classes, such as hormones (which govern functions like appetite, metabolism, growth, reproduction and sleep) and neurotransmitters (which govern an almost endless range of neurological functions). But scientists are discovering that chemical messengers have an almost universal spectrum of functions that converge in the immune system.

This makes sense, because the immune system really isn't about immunity at all, but about integrity. It is not surprising that the endocrine system and nervous system converge in the system responsible for maintaining integrity.

The immune system includes billions of cells that circulate throughout the body, picking up and distributing information. Some immune cells circulate, some remain in fixed positions and some carry signals between the circulating cells and the fixed cells.

Circulating immune cells are called white blood cells (WBCs) or leukocytes. Fixed immune cells have different names, depending on their location. For example, immune cells in the liver are called Kupfer cells, while similar cells in the brain are called glial cells.

antibodies. Once these invaders are identified, GALT acts swiftly to neutralize them so they cannot cause harm. In its other immune role, GALT is involved in discharging waste products and promoting repair and recovery of injured mucosal tissue to restore the immunological integrity of the intestinal mucosal barrier.

Who's Minding the House?

Every time we eat, drink or breathe, harmful substances can enter the body. For example, the salad you ate with dinner was wonderfully nutritious, but did the lettuce leaves harbor any molds or spores? Did the avocado pick up a few bacteria from the cutting board? Did airborne microbes settle on your glass of water and enter your body when you took a sip?

The answer is yes. This kind of thing happens all of the time. We try to protect children from germs, but the human body has been exposed to the environment throughout its entire history—long before the invention of soap and antibiotics. If humans didn't have robust immune systems, we never would have survived. An essential phase of immune system development takes place *after* birth, when infants and young children are exposed to potential dangers in their environment. With each exposure—including some that may cause illness—the immune system learns to recognize dangers and to remember them for the future.

This is where friendly bacteria come in. One of the most important things that friendly bacteria do in the body is completely passive—they house sit. Think of beneficial intestinal bacteria as a permanent flotilla of friendly housesitting neighbors; they simply occupy the terrain, implanting themselves tightly into the tissues of the intestines and literally becoming a part of the digestive system.

Cutting boards can harbor bacteria, which are easily transferred to food.

When other microbes—like harmful bacteria, molds and yeasts—enter the digestive tract, they are met by overwhelming numbers of the friendly bacteria that live there. If the new organisms try to permanently establish occupancy, they find a big "no vacancy" sign. If the microbes are aggressive and try to attack, the immune system is prepared to resist. As long as the intestinal flora is healthy, it's extremely difficult for foreign microbes to vandalize the house. Only microorganisms that are prepared to be part of a peaceful community are welcome.

Favoring Our Friends

Maintaining a healthy population of beneficial bacteria is vital for overall health and well-being. When healthy bacteria are in the digestive tract, harmful bacteria have less opportunity to take control and wreak havoc on the body. If you've ever suffered from an intestinal virus or food poisoning, then you know just how miserable you can feel when the balance tips in favor of harmful microorganisms.

Ironically, one of the most devastating influences on healthy intestinal flora—and therefore, on the ability of the immune system to keep us healthy—is the excessive or improper use of antibiotic drugs. Antibiotic drugs are compounds that destroy or inhibit the growth of bacteria. Without a doubt, antibiotics are lifesavers. In the nearly 80 years since the discovery of penicillin, antibiotics have saved countless lives that would otherwise have been lost to serious infectious illnesses. Antibiotic drugs changed the face of modern medicine.

But there is great danger. All too often, doctors have prescribed antibiotics too freely. Antibiotics have been used to suppress simple infections that, if allowed to run their course, would be resolved naturally by the immune system. For example, until recently, children with mild ear infections would almost inevitably receive prescriptions for antibiotics.

Antibiotics also have been prescribed in countless instances for viral infections (a useless practice, since antibiotics do not kill viruses). There are several important reasons why these and other misuses of antibiotics present a problem.

Antibiotics don't only kill pathogenic (disease-causing) bacteria; they also kill friendly bacteria. The loss of friendly bacteria means that nutrients are not assimilated as well. Production of vitamins and other nutrients is diminished, which exerts an impact on many body functions, including the formation of healthy blood cells, the proper replication and repair of DNA (deoxyribonucleic acid) and many aspects of immune system communication.

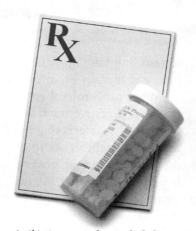

Antibiotic use can disrupt the balance of flora in the digestive tract.

When friendly bacteria are destroyed, their ability to house sit is diminished. Antibiotics—even when prescribed appropriately—make humans more susceptible to intestinal infections. This is especially true when a prescribed course of antibiotics is not completed (as is often the case). Many people stop taking their antibiotics as soon as they feel better rather than completing the entire seven- to ten-day course. In such a case, pathogenic bacteria remain in the body in numbers too small to cause discomfort, but they can rebound. Freed from the lethal pressure of antibiotics, pathogenic bacteria are able to reproduce in large numbers. Since friendly, housesitting bacteria may have also been killed off, it is easier for pathogenic bacteria to move in.

While antibiotics kill most harmful bacteria, a small number of bacteria may acquire antibiotic-resistant genes *(see sidebar—"When Good Germs Go Bad")*. If antibiotic-resistant bacteria have a chance to proliferate, they can cause an infection worse than the one for which the antibiotics were originally prescribed.

This effect—which results in part from the overuse of antibiotics—is one reason why life-threatening, super-resistant bacterial strains are developing, forcing pharmaceutical companies to produce more powerful drugs. It is also a reason that some nearly eradicated bacterial illnesses are returning and posing serious health threats.

Overuse of antibiotic drugs may prompt bacteria to morph in potentially dangerous ways. Bacteria are highly adaptive life forms. Most—if not all—bacteria are pleomorphic, meaning that they can take on several forms. Bacteria choose among their different forms according to the cues they receive from the environment. For example, when nutrients become scarce, some bacteria shift into a dormant form that consumes almost no energy. They can wait in this state for thousands of years until they are rehydrated and supplied with nutrients.

When antibiotics are introduced into the environment, many bacteria shift into a CWD (cell wall deficient) variant. This means the bacterium sheds its complex cell wall in favor of a simple membrane. Unfortunately, the cell wall contains most of the markers the immune system uses to identify harmful bacteria.

In their CWD form, bacteria effectively drop off the immune system's radar screen like high-tech stealth bombers. Undetected by the immune system, CWD bacteria can move deep into the body, setting up residence inside cells, nerve tissues and other places where it is difficult for the immune system to find them and root them out.

In some cases, the immune system can sense the presence of an unwanted microbe but is unable to locate the invader. In such cases, the immune system may initiate an inflammatory reaction—a kind of biochemical brush fire—in an attempt to destroy invaders it cannot see. Some scientists now believe that such crude attempts to attack invisible pathogens are responsible for many inflammatory disorders, including autoimmune diseases and atherosclerosis (the buildup of arterial plaque).

Intestinal Conditions and Diseases

The optimal balance of beneficial bacteria is known as symbiosis. The condition in which intestinal flora is out of balance is known as dysbiosis. There are four general causes and types of dysbiosis:

Putrefaction dysbiosis. High-fat diets, especially those high in animal fat and low in roughage or insoluble fiber, can cause putrefaction dysbiosis. (Sources of insoluble fiber include green leafy vegetables, whole fruits and unrefined foods such as bran and sprouted seeds.)

Fermentation dysbiosis. Bacterial overgrowth in the digestive tract can cause fermentation dysbiosis. When this overgrowth occurs, dietary carbohydrates ferment in the digestive tract and produce toxic waste products. For example, fermented sugar produces ethanol, which over time can damage the intestines.

Deficiency dysbiosis. Deficiency dysbiosis occurs when too few beneficial bacteria line the digestive tract. Antibiotics are often the culprits, because they do not discriminate between beneficial and harmful bacteria—they simply wipe out a substantial portion of the intestinal flora. A diet low in soluble fiber also may cause deficiency dysbiosis. (Sources of soluble fiber include oat bran, barley, nuts, seeds, beans and some fruits and vegetables.)

Sensitization dysbiosis. Sensitization, or altered immune responses to intestinal flora, may contribute to a variety of disorders, including autoimmune disorders. Such disorders include irritable bowel disease, Crohn's disease, ulcerative colitis and some types of arthritis, connective tissue diseases and skin diseases.

Because GALT and healthy intestinal flora are so important for maintaining good health, many problematic health conditions are linked to dysbiosis and intestinal health. Some of these conditions are digestive disorders; others are related to abnormal immune responses. Dysbiosis may interfere with nutrient absorption, resulting in nutritional deficiencies, which in turn can produce other health problems.

Dysbiosis also may lead to an overgrowth of fungi, yeast or harmful bacteria. When harmful bacteria predominate in the intestinal flora, their enzymes compromise digestion and the integrity of the lining of the digestive tract. The most common type of bacterial dysbiosis, candidiasis, is caused by *candida albicans*, a form of yeast.

Dysbiosis also leads to food sensitivities and allergies by causing leaky gut syndrome (also known as intestinal permeability). Leaky gut syndrome occurs when the tight network of cells lining the digestive tract loosens, allowing undigested food particles to leak into surrounding tissues. Since these tissues are unprepared to contend with food particles, food intolerance and allergies may develop. This can prompt a vicious cycle in which food allergies increase leaks in the gut, which in turn aggravates food allergies. (Research suggests that stress plays a role in leaky gut syndrome and other intestinal disorders.)

Dysbiosis can also cause impaired immune cells to leave the digestive tract and travel to other sites in the body, where they release chemicals that cause inflammation. Some types of arthritis have been linked to bacterial infections in the intestines.

Dysbiosis has been recognized as a contributing factor or underlying cause of the following disorders:

Acne	Eczema and psoriasis
Allergies	Endometriosis
Anorexia and bulimia	Fibrositis
Arthritis	Gastritis
Asthma	Headache
Attention deficit disorder	Hormonal disturbances
Candida	Hypoglycemia
Cancer	(low blood sugar)
Chronic fatigue	Irritable bowel disease
Colitis	Menstrual disorders
Constipation	Muscle pain
Diarrhea	Obesity
Crohn's disease	Vaginal infections
Depression	

When Is an Allergen Not an Allergen?

For any given person, a substance typically is or isn't an allergen—that is, you're allergic to something or you're not. But different parts of the digestive system view things differently.

For example, many people are allergic or sensitive to gluten, a protein that is a natural component of wheat. Gluten-intolerant people often experience relief from digestive problems when they eliminate bread, pasta and other high-gluten foods from their diet. But consider the following: it's possible to remove gluten from wheat flour, leaving behind flour that makes very dull bread. As expected, people with gluten intolerance do not react badly to bread made with this flour (except to complain about its terrible texture).

Suppose you remove the gluten from flour, but put it back in before you make the bread? You'd expect gluten-intolerant people to have a bad reaction, right? But many gluten-intolerant people feel fine when they eat bread made this way. What is happening?

Removing and re-adding gluten to bread may sometimes prevent gluten intolerance.

The answer is simple. In natural flour, gluten molecules are bundled up deep inside strands of starch (long chains of sugar molecules). When we eat bread, the digestive system tears apart the starch molecules. If our ability to digest starch is impaired for some reason, we may not break down enough starch to expose the gluten protein. Consequently, the protein isn't digested and broken down into components that the lower portions of the digestive system can recognize and handle. When the intact protein molecule enters the lower intestinal tract, the immune system tags it as an invading substance. If the starch surrounding the gluten had been fully digested in the upper digestive tract, then the gluten also would have been exposed and digested.

When gluten is removed from flour and mixed back in before the bread is baked, the gluten molecules are no longer inside the starch molecules. Even if the starch is not fully digested, the gluten is exposed to the digestive system. Proteolytic (protein-breaking) enzymes immediately process the protein, breaking it down into individual amino acids and other simpler molecules. In the lower digestive tract, these molecules are recognized as nutrients—not invaders.

For the immune system to work best, we need to digest our foods fully. This not only provides us with access to all the nutrients in the food, but also prevents the immune system from generating false alarms. When the immune system isn't dealing with false alarms, it can better respond to real threats and challenges.

Acidity and Digestive Health

A high fat meal may cause some people to experience indigestion.

Carbohydrate digestion requires proper stomach acidity. A healthy stomach is between 100,000 and 1,000,000 times more acidic than water. However, many people are deficient in stomach acid, which may allow food particles to enter the lower intestinal tract in forms that the immune system may recognize as a potential threat.

This is why some people suffer from indigestion. Indigestion results from incomplete digestion of meals—especially high-fat meals—and causes temporary abdominal discomfort, fullness, gas and bloating.

As the acidic contents of the stomach enter the small intestine, they must be immediately cooled down with an alkalizing bath of sodium bicarbonate provided by the pancreas (the pancreas secretes various enzymes into the intestine to aid in the digestion process). When the pancreas is not functioning properly, there may be unpleasant symptoms. Indigestion caused by pancreatic insufficiency can generally be resolved by supplemental pancreatic enzymes.

Many digestive enzymes are highly dependent on pH (a measure of acidity and alkalinity). If food entering the intestines remains too acidic, it will not be properly digested and will pass through the digestive tract in a potentially allergy-inducing form.

When pH values are chronically out of balance, it becomes very difficult for friendly bacteria to survive. Potentially harmful bacteria tend to thrive, as do molds and yeasts that prefer a more acidic environment, displacing beneficial bacteria and leading to a breakdown of digestive and intestinal immune function.

The digestive processes and their coordination with the immune system are complex, and there are many ways they may become derailed. Fortunately, each difficulty also reveals an opportunity to enhance digestion and intestinal immunity through nutrition, supplementation and lifestyle changes.

Supporting Digestive Health

Stress management and a healthy diet can enhance digestion and intestinal immunity. Many people may also benefit from supplementing their diet with probiotics, prebiotics and digestive enzymes to favor the growth of beneficial intestinal flora.

Probiotics

Probiotics, which means, "for life," are products that contain live microorganisms in sufficient numbers to alter the intestinal microflora, promote intestinal microbial balance and benefit health.

Before the advent of refrigeration and processed foods, people routinely consumed live bacteria and other microorganisms in fermented foods. Today, there are still food sources of probiotic bacteria—some examples are yogurt, kimchee and sauerkraut—but few Americans consume these foods regularly. Probiotics are also available as liquids, powders and pills and have been used for many years without any reported ill effects.

In the early twentieth century, researchers observed that probiotics benefited the health and lifespan of Bulgarian peasants who consumed fermented milk products such as yogurt and kefir. Probiotics improve the microbial balance in the digestive tract and enhance overall health and digestion. Probiotics promote digestion

Yogurt can be a potent source of live, beneficial microorganisms.

and uptake of nutrients by the intestine. They also help regulate immune function and strengthen resistance to pathogenic bacteria. Numerous clinical trials have shown probiotics to support a fully functional digestive system and a robust intestinal immunity.

There are many beneficial bacteria, but among the most important are *lactobacilli* and *bifidobacteria*, both of which have been extensively studied for their health benefits. These bacteria are linked to protection from disease-causing microorganisms, relief of lactose intolerance, relief from digestive disorders (such as constipation), lower cholesterol, increased immune function and other health benefits.

Probiotics protect against harmful bacteria by penetrating and binding to the surfaces of harmful organisms; stimulating the lining of the digestive tract to prevent penetration by pathogens; and modifying immunoregulation—decreasing molecules that may cause harm and increasing those that offer protection.

Probiotics have been shown to relieve temporary abdominal bloating and to reduce intestinal gas. In a randomized, double blind study (the most rigorous study design in which neither subjects nor researchers know whether they are receiving, or administering the study agent or placebo), subjects were given either a *Lactobacillus rhamnosus* supplement or a placebo. The researchers found that the probiotic supplement significantly improved symptoms of temporary abdominal bloating.

There also is mounting evidence probiotics exert other powerful health benefits, including the following:

- Reducing infections (especially *Helicobacter pylori* infection, which is associated with ulcers)
- Reducing allergic symptoms
- Relieving constipation and diarrhea
- Relieving symptoms of irritable bowel syndrome
- Reducing cholesterol and triglyceride levels
- Enhancing mineral metabolism

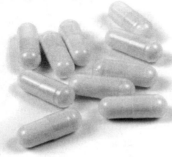
Probiotic formulas are available in powder or capsule form.

- Improving bone density and stability

Because supplemental probiotics do not permanently repopulate the intestine, you must ingest them regularly to achieve lasting health benefits. Since passage through the digestive system can reduce the number of probiotic bacteria that reach the intestines in a live, active state, it's generally recommended that probiotic supplements be taken several hours before or after eating, when the acidity of the stomach is at its lowest. Some probiotic formulas use special coatings or pearl packaging to prevent the delicate bacteria from being released until they have passed through the stomach and reach the duodenum, the first segment of the intestinal tract.

Many popular probiotic formulas combine several different strains of probiotic bacteria—often as many as eight or more in a single capsule. Supplementing with several different strains may provide a broader spectrum of benefits, as certain strains may be more successful at colonizing and reproducing in different parts of the digestive tract under different conditions such as pH (acid/base balance).

Antibiotic drugs, frequently prescribed to help control infection by disease-causing bacteria, can kill off a portion of our body's natural bacterial flora. Since these bacteria are essential in supporting important immune functions, it's important to help reestablish a healthy intestinal ecosystem as soon as possible following drug use. Some antibiotic drugs are designed and chosen to minimize their negative impact on our beneficial flora. Conversely, some strains of probiotic bacteria have been bred to be particularly resistant to antibiotic medicines.

Still, following a course of either prescription antibiotics or any herbs and natural supplements that are highly bactericidal ("bacteria killing"), it's especially important to work to repopulate the body's natural bacterial ecosystem. Some naturally bactericidal substances include oregano oil, tea tree oil, grapefruit seed extract and olive leaf extract. Garlic is often cited as a natural antibiotic, but current research suggests that garlic seems to help by interfering with the signals bacteria use to communicate in the formation of their disease causing colonies, a property called *quorum sensing*, rather than through a strong bacteria killing effect.

There is no general agreement on whether it's beneficial to take probiotics at the same time that one is taking a course of antibiotic medicine. A scientifically accurate answer is likely dependent upon the individual's unique bacterial ecosystem, the antibiotics involved and the specific strains of probiotic bacteria. Since there is no simple formula, some practitioners suggest aggressively supplementing with probiotics *immediately after* the last dose of antibiotic medication and continuing for *at least* one month to six weeks.

Prebiotics

Prebiotics are nondigestible food ingredients that stimulate the growth and activity of certain bacteria in the colon. Prebiotics are primarily oligosaccharides—short chains of sugar molecules that can only partially be digested. Most naturally occurring oligosaccharides are found in plants. Prebiotics are often called bifidogenic factors because they primarily stimulate growth of bifidobacteria.

Prebiotics work in concert with probiotics and have been found to produce many of the same health benefits. There is also evidence that prebiotics play a role in cancer prevention by binding and inactivating certain cancer-causing agents, directly inhibiting the growth of some tumors and inhibiting bacteria associated with cancer-causing substances. In addition, prebiotics help lower blood lipids—cholesterol and triglycerides—and modulate blood sugar.

While some probiotic supplements are formulated to include prebiotic compounds, it's also possible to purchase prebiotics separately, as capsules and powders. FOS—short for Fructooligosaccharide—is one popular alternative. FOS is a naturally sweet polymer that links together several molecules of the simple sugars glucose and fructose. Although it's about half as sweet as table sugar, it is not recognized or metabolized by the body and therefore is practically non-caloric and non-glycemic. (Unlike table sugar, FOS stimulates almost no insulin response). FOS and similar compounds are classified as soluble fiber and are prized for their beneficial impact on intestinal health.

Inulin is another type of soluble fiber and is primarily a polymer of fructose. It, too, can have a beneficial prebiotic function, but inulin can sometimes trigger the development of gas in the large intestine as bacteria break it down and produce methane, hydrogen and carbon dioxide. When supplementing with inulin, many health professionals suggest starting with a small amount and adjusting the amount based on how your body responds.

Digestive Enzymes

Digestive enzymes allow us to use the nutrients in food. The pancreas produces the major digestive enzymes that break down each type of nutrient: protease, which breaks proteins down into amino acids; lipase, which breaks fats down into fatty acids; and amylase, which breaks complex carbohydrates down into simple sugars. These three key enzymes are important for healthy digestion and support the health and strength of intestinal immunity.

Raw foods naturally contain enzymes that have not been deactivated by cooking.

Raw foods generally contain active enzymes that naturally aid in their digestion. However, the more cooked and processed foods we eat, the more we rely on the body's ability to manufacture digestive enzymes. Over time, the continued production of enzymes can become quite taxing, and at some point—especially under conditions of stress or illness—we may not produce adequate quantities of some critical enzyme groups. This can lead to incomplete assimilation of critical nutrients from our food, slower transit times through the digestive system, shifts in the critical acid/base balance in the blood, tissues and digestive system, and changes in the bacterial ecosystem within the intestines.

Over-the-counter and specialty digestive enzyme formulations generally contain a spectrum of different types of enzymes—not only three basic classes listed above, but also different forms of these enzymes that are active under different chemical conditions and can operate in different parts of the digestive tract.

These supplements are best taken at the beginning of a meal so they can work in tandem with the body's natural digestive processes to break down the foods we eat. However, they can be effective even when taken during or a short while after a meal.

Some specialized enzymes, such as high potency proteolytic (protein breaking) formulas available from healthcare professionals, are sometimes taken on an empty stomach rather than with a meal. These can help break down excess protein wastes that can build up in the blood and lymphatic systems.

Some Food for Thought

The human digestive system is a miracle of under-appreciated complexity. We can eat practically anything that grows and our bodies will figure out how to process, use and discharge it. Because it's so versatile, it's easy to take the digestive system for granted.

But besides extracting energy from food to keep us alive with the chemical building blocks we need to grow and repair our bodies, GALT is the true frontier of the immune system, not only recognizing and assimilating critical nutrients, but also analyzing and protecting us from an ever-shifting array of disease-causing organisms and toxins. In addition to its innate, built-in intelligence, the GALT is always learning, constantly adapting to new threats in an endless quest to keep us healthy and whole. Therefore, it's vitally important to keep GALT healthy so that it can do its job and keep us healthy.

Natural Support for Digestion

Aloe naturally soothes digestive discomfort.
Anise, a member of the parsley family, supports digestion and relieves gas.
Cinnamon improves digestion, helps control diarrhea and relieves abdominal spasms due to gas.
Dandelion greens stimulate healthy intestinal function.
Fennel increases digestion, relieves gas and aids in proper waste elimination.
Ginger can help to relieve discomfort by supporting intestinal function and is one of the best herbs for nausea.
Lemon peel has been used for nausea and abdominal bloating.
Peppermint tea may relieve bloating and nausea.
Turmeric may help with gas and indigestion.

Test Your GI IQ!

The answers to these and other questions about the digestive tract can be found in this booklet. Test yourself before and after reading *Digestive Health*.

1. The digestive tract includes multiple organs in the body. Which organ is NOT part of the digestive tract?
    a. The esophagus
    b. The thyroid
    c. The colon
    d. The stomach
2. How many friendly bacteria are in healthy intestines?
    a. Thousands
    b. Millions
    c. Billions
    d. Trillions
3. Which of the following are symptoms of digestive disorders?
    a. Indigestion
    b. Gas
    c. Bloating
    d. All of the above

4. When should you stop taking antibiotics?
    a. When you finish the prescribed amount
    b. When you start to feel better
    c. When the symptoms completely disappear
5. Which of the following occurs when too few bacteria line the intestinal tract?
    a. Putrefaction symbiosis
    b. Fermentation dysbiosis
    c. Deficiency dysbiosis
6. Which of the following factors causes indigestion?
    a. Incomplete digestion of meals
    b. Overcooked foods
    c. A carbohydrate meal
7. Which organ produces the most digestive enzymes (substances that help break down food)?
    a. The colon
    b. The pancreas
    c. The kidneys
8. Which of the following statements correctly describes GALT?
    a. Specific components include the tonsils, adenoids, and Peyer's patches
    b. A network of immune system cells located throughout the digestive tract
    c. Necessary to keep us healthy.
    d. All of the above
9. Is the following statement true or false? Heartburn is caused by excessive acid.
    a. True
    b. False
    c. Neither
10. A healthy stomach is how many times more acidic than water?
    a. 1–100,000
    b. 10,000–50,000
    c. 100,000–1,000,000

Answers: 1.B, 2.D, 3.D, 4.A, 5.C, 6.A, 7.B, 8.D, 9.B, 10.C

References

Aattour, N. et al. 2002. "Oral ingestion of lactic-acid bacteria by rats increases lymphocyte proliferation and interferon-gamma production." *British Journal of Nutrition* 87(4): 367–73.

Bertazzoni, M.E. et al. 2001. "Preliminary screening of health-promoting properties of new lactobacillus strain: in vitro and in vivo." *HEALFO Abstracts*.

Casas, I.A. and W.J. Dobrogosz. 2000. "Validation of the probiotic concept: *Lactobacillus reuteri* confers broad-spectrum protection against disease in humans and animals." *Microbial Ecology in Health and Disease* 12(4): 247–85.

Collins, M.D. and G.R. Gibson. 1999. "Probiotics, prebiotics, and synbiotics: approaches for modulating the microbial ecology of the gut." *American Journal of Clinical Nutrition* 69(5): 1052S–1057S.

Delzenne, N.M. and C.M. Williams. 2002. "Prebiotics and lipid metabolism." *Current Opinion in Lipidology*. 13(1): 61–67.

di Stefano, M. et al. 2004. "Probiotics and functional abdominal bloating." *Journal of Clinical Gastroenterology* 38(2): S102–03.

Dodd, H.M. and M.J. Gasson. Bacteriocins of Lactic Acid Bacteria. In *Genetics and biotechnology of lactic acid bacteria*, ed. M.J. Gasson and W.M. de Vos, 211–51. Glasgow, UK: Blackie Academic and Professional.

Dowell, S.F. et al. 1998. "Otitis media—principles of judicious use of antimicrobial agents." *Pediatrics* 101(1): 165–71.

D'Souza, A.L. et al. 2002. "Probiotics in prevention of antibiotic-associated diarrhoea: meta-analysis." *BMJ* 324(7350): 1361.

Elmer, G.W. 2001. "Probiotics: living drugs." *American Journal of Health-System Pharmacy* 58(12): 1101–09.

Isolauri, E. et al. 2002. "Probiotics: a role in the treatment of intestinal infection and inflammation?" *Gut*(Suppl 3): III54–59.

Jiang, T. et al. 1996. "Improvement of lactose digestion in humans by ingestion of unfermented milk containing *Bifidobacterium longum*." *Journal of Dairy Science* 79(5): 750–57.

Kailasapathy, K. and J. Chin. 2000. "Survival and therapeutic potential of probiotic organisms with reference to *Lactobacillus acidophilus* and *Bifidobacterium* spp." *Immunology and Cell Biology* 78(1): 80–88.

Lebenthal, E. et al. 1993. "Pancreatic extract lipase activity." *JAMA* 270(21): 2557–58.

Macfarlane, G.T. and J.H. Cummings. 1999. "Probiotics and prebiotics: can regulating the activities of intestinal bacteria benefit health? *BMJ* 318(7189): 999–1003.

Mattman, L.H. 2001. *Cell Wall Deficient Forms: Stealth Pathogens*, 3rd ed. Boca Raton, FL: CRC Press.

McLaughlin, R.W. et al. 2002. "Are there naturally occurring pleomorphic bacteria in the blood of healthy humans?" *Journal of Clinical Microbiology* 40(12): 4771–75.

Rakel, R.E. and E.T. Bope. 2005. *Conn's Current Therapy 2005*, 57th ed. Philadelphia, PA: Saunders Elsevier.

Sartor, R.B. 2005. "Probiotic therapy of intestinal inflammation and infections." *Current Opinion in Gastroenterology* 21(1): 44–50.

Schrezenmeir, J. and M. de Vrese. 2001. "Probiotics, prebiotics, and synbiotics—approaching a definition." *American Journal of Clinical Nutrition.* 73(2 Suppl): 361S–364S.

Söderholm, J.D. and M.H. Perdue. 2001. "Stress and gastrointestinal tract. II. Stress and intestinal barrier function." *American Journal of Physiology, Gastrointestinal and Liver Physiology* 280(1): G7–G13.

Suarez, F. et al. 1999. "Pancreatic supplements reduce symptomatic response of healthy subjects to a high-fat meal." *Digestive Diseases and Sciences* 44(7): 1317–21.

Vanderhoof, J.A. and R.J. Young. 2004. "Current and potential uses of probiotics." *Annals of Allergy, Asthma and Immunology* 93(5 Suppl 3): S33–S37.

Walker, A.W. and L.C. Duffy. 1998. "Diet and bacterial colonization: role of probiotics and prebiotics." *Journal of Nutritional Biochemistry* 9(12): 668–75.

Yan, F. and D.B. Polk. 2004. "Commensal bacteria in the gut: learning who our friends are." *Current Opinion in Gastroenterology* 20(6): 565–71.

Check out these other top-selling Woodland Health Series booklets:

Ask for them by title or ISBN at your neighborhood bookstore or health food store. Call Woodland at (800) 777-BOOK for the store nearest you.

Açaí Berry	978-1-58054-472-6
Alzheimer Disease: A Naturopathic Approach	978-1-58054-423-8
Bee Pollen (2nd ed.)	978-1-58054-429-0
Candida Albicans (2nd ed.)	978-1-58054-432-0
Chelation Therapy (2nd ed.)	978-1-58054-431-3
Chinese Red Yeast Rice (2nd ed.)	978-1-58054-434-4
Coconut Oil	978-1-58054-464-1
Coenzyme Q10	978-1-58054-456-6
Colon Health (2nd ed.)	978-1-58054-435-1
Conjugated Linoleic Acid (2nd ed.)	978-1-58054-433-7
Cranberries	978-1-58054-461-0
Digestive Enzymes (2nd ed.)	978-1-58054-436-8
Fertility: A Naturopathic Approach	978-1-58054-466-5
Fish Oil, Omega-3 and Essential Fatty Acids (2nd ed.)	978-1-58054-437-5
Flaxseed Oil (2nd ed.)	978-1-58054-438-2
Ginkgo Biloba	978-1-88567-010-6
Ginseng	978-1-58054-483-2
Goji Berry	978-1-58054-473-3
Grapefruit Seed Extract (2nd ed.)	978-1-58054-446-7
HPV and Cervical Dysplasia	978-1-58054-463-4
Hoodia (2nd ed.)	978-1-58054-448-1
Hyaluronic Acid	978-1-58054-458-0

Influenza, Epidemics, Bird Flu	978-1-58054-425-2
Insomnia	978-1-58054-117-6
Liver Health	978-1-58054-397-2
Managing Acid Reflux (2nd ed.)	978-1-58054-444-3
Mangosteen	978-1-58054-470-2
Nattokinase	978-1-58054-172-5
Natural Guide to Back Pain	978-1-58054-457-3
Natural Guide to Energy Enhancers	978-1-58054-414-6
Natural Guide to Managing Pre-Diabetes	978-1-58054-465-8
Natural Guide to Relieving Headache Pain	978-1-58054-443-6
Olive Leaf Extract (2nd ed.)	978-1-58054-441-2
Oregano Oil (2nd ed.)	978-1-58054-475-7
Pomegranate	978-1-58054-471-9
Stevia (2nd ed.)	978-1-58054-476-4
Stress (2nd ed.)	978-1-58054-477-1
Supplements for Fibromyalgia	978-1-58054-034-6
Xylitol	978-1-58054-139-8

Each booklet is approximately 32 pages and priced at $4.95.

Healthy Reading for More Than 30 Years.

About the Author

Barbara Wexler is a medical writer and chronic disease epidemiologist who brings more than 25 years of experience as a clinician, researcher, educator and administrator to the articles and texts she prepares for professional and consumer audiences. A graduate of Sarah Lawrence College and the Yale University College of Medicine, School of Epidemiology and Public Health, Wexler is interested in evidence-based complementary and integrative medicine.